SpringerBriefs in Applied Sciences and Technology

Series editor

Andreas Öchsner, Griffith School of Engineering, Griffith University, Queensland, Australia

SpringerBriefs present concise summaries of cutting-edge research and practical applications across a wide spectrum of fields. Featuring compact volumes of 50–125 pages, the series covers a range of content from professional to academic.

Typical publications can be:

- A timely report of state-of-the art methods
- An introduction to or a manual for the application of mathematical or computer techniques
- A bridge between new research results, as published in journal articles
- A snapshot of a hot or emerging topic
- An in-depth case study
- A presentation of core concepts that students must understand in order to make independent contributions

SpringerBriefs are characterized by fast, global electronic dissemination, standard publishing contracts, standardized manuscript preparation and formatting guidelines, and expedited production schedules.

On the one hand, **SpringerBriefs in Applied Sciences and Technology** are devoted to the publication of fundamentals and applications within the different classical engineering disciplines as well as in interdisciplinary fields that recently emerged between these areas. On the other hand, as the boundary separating fundamental research and applied technology is more and more dissolving, this series is particularly open to trans-disciplinary topics between fundamental science and engineering.

Indexed by EI-Compendex and Springerlink.

More information about this series at http://www.springer.com/series/8884

P. Mahima · M. Suprava
S. Vandana · Mohammed P. S. Yazeen
Raveendranath U. Nair

Electromagnetic Performance Analysis of Graded Dielectric Inhomogeneous Radomes

 Springer

P. Mahima
Centre for Electromagnetics (CEM)
CSIR-National Aerospace Laboratories
 (CSIR-NAL)
Bangalore
India

M. Suprava
Centre for Electromagnetics (CEM)
CSIR-National Aerospace Laboratories
 (CSIR-NAL)
Bangalore
India

S. Vandana
Centre for Electromagnetics (CEM)
CSIR-National Aerospace Laboratories
 (CSIR-NAL)
Bangalore
India

Mohammed P. S. Yazeen
Centre for Electromagnetics (CEM)
CSIR-National Aerospace Laboratories
 (CSIR-NAL)
Bangalore
India

Raveendranath U. Nair
Centre for Electromagnetics (CEM)
CSIR-National Aerospace Laboratories
 (CSIR-NAL)
Bangalore
India

ISSN 2191-530X   ISSN 2191-5318   (electronic)
SpringerBriefs in Applied Sciences and Technology
ISBN 978-981-10-7831-6   ISBN 978-981-10-7832-3   (eBook)
https://doi.org/10.1007/978-981-10-7832-3

Library of Congress Control Number: 2017964578

Printed on acid-free paper

This Springer imprint is published by Springer Nature
The registered company is Springer Nature Singapore Pte Ltd.
The registered company address is: 152 Beach Road, #21-01/04 Gateway East, Singapore 189721, Singapore

Preface

The advancements in airborne radar systems demand novel airborne radome designs. This necessitates the electromagnetic design of new radome wall structures (based on FSS, inhomogeneous dielectric structures, etc.) to meet the stringent airborne radome performance requirements. In view of this, the design and EM performance analysis of a graded dielectric inhomogeneous radome wall is presented in this brief. Various performance parameters like co-pol power, cross-pol power, boresight error are analyzed, and their dependence on material parameters of the radome wall is investigated.

The work in this brief is organized as follows: Chap. 1 deals with the introduction of radomes, and Chap. 2 gives us information on various radome wall configurations. The characteristics of a graded dielectric radome wall structure are covered in Chap. 3. Here, planar slab model of the wall structure is considered. Further, the design of the graded dielectric radome wall based on equivalent transmission line method is presented in Chap. 4 which includes designing for two radome shapes, hemispherical and tangent ogive. An elaborate analysis of the EM characteristics of the antenna-radome system based on geometrical optics method in conjunction with aperture integration method is discussed in Chap. 5 that gives an idea of the effect of radome on the antenna characteristics. The last chapter gives us the summary of the whole work done.

Bangalore, India

P. Mahima
M. Suprava
S. Vandana
Mohammed P. S. Yazeen
Raveendranath U. Nair

Acknowledgements

We would like to thank Mr. Shyam Chetty, Director, CSIR-National Aerospace Laboratories, Bangalore, for giving permission to write this Springer Brief.

Further, we record our thanks to colleagues Dr. Shiv Narayan, Dr. Hema Singh, Dr. Balamati Choudhury, and Mr. K. S. Venu for their suggestions and cooperations. We express our sincere thanks to Project Staff of Centre for Electromagnetics, Ms. Vinisha C. V., Ms. Karthika S. Nair, Ms. Maumita Dutta, for their consistent support during the preparation of the technical contents of this book.

We express our gratitude to Dr. Suvira Srivastav, Associate Director, Springer (India) Private Limited, for giving us the opportunity to write this book. Nevertheless, it would not have been possible to bring out this book within a short span of time without consistent support and suggestions of Dr. Swati Mehershi, Senior Editor, Applied Sciences and Engineering, Springer India. We highly appreciate the effort made by Ms. Kamiya Khattar, Senior Editorial Assistant—Hard Sciences—and Ms. Aparajita Singh, Editorial Assistant—Physical Sciences and Engineering—Springer India to accomplish this book.

About the Book

The state-of-the-art radomes exhibit less operational bandwidth. Development of high-performance radomes for both airborne and ground-based applications requires a novel radome wall structure that provides minimal degradation to the antenna radiation pattern for a wide range of frequencies and is easy to fabricate. This can be attained by designing a radome wall with varying dielectric parameters. This book presents a novel approach in which multilayered radome wall based on inhomogeneous planar layer is designed.

In this book, the design of graded dielectric radome with seven layers with a varying thickness of different dielectric constant and loss tangent is presented in detail. The EM design of proposed radome wall has been carried out using equivalent transmission line method. The radome wall exhibits excellent transmission characteristics. The EM performance analysis of antenna-radome system with graded dielectric radome wall is performed by geometrical optics method in conjunction with aperture integration method. The performance parameters of the antenna-radome system have been computed and presented in this book which shows the proposed radome wall is an excellent candidate for radome applications.

Contents

About the Authors

P. Mahima is currently working as a Project Scientist in Centre for Electromagnetics (CEM), CSIR-National Aerospace Laboratories (CSIR-NAL), Bangalore, India, since June 2016. She was born in 1991 in Kerala, India. She obtained her M.Tech. degree in Microwave and Radar Electronics from Cochin University of Science and Technology (CUSAT). She completed her B.Tech. degree in Electronics and Communication Engineering from Nehru College of Engineering and Research Centre, Pampady, Kerala, India, in 2010. Her research interests are in design and analysis of radomes and antennas.

M. Suprava was working as a Project Postgraduate Trainee in Centre for Electromagnetics (CEM), CSIR-National Aerospace Laboratories (CSIR-NAL), Bangalore, India, from August 2014 to November 2015. She was born in 1993 in Odisha, India. She obtained her B.Tech. degree in Electronics and Communication Engineering from Vignan Institute of Technology and Management Engineering, Brahmapur, Odisha, India, in 2010. Her research interests are in design and analysis of radomes.

S. Vandana was working as a Project Postgraduate Trainee in Centre for Electromagnetics (CEM), CSIR-National Aerospace Laboratories (CSIR-NAL), Bangalore, India, from October 2014 to July 2015. She was born in 1988 at Kerala, India. She obtained her B.Tech. degree in Electronics and Communication Engineering from College of Engineering, Kalloopara, Kerala, India, in 2010 and M.Tech. degree in Electronics with specialization in Digital Electronics from Department of Electronics, Cochin University of Science and Technology (CUSAT), Cochin, India, in 2013. Her research interests are design and analysis of radomes, signal processing, and EM material characterization techniques.

Mohammed P. S. Yazeen is currently working as a Project Scientist in Centre for Electromagnetics (CEM), CSIR-National Aerospace Laboratories (CSIR-NAL), Bangalore, India, since February 2015. He was born in 1990 in Kerala, India. He obtained his M.Tech. degree in Electronics with specialization in Wireless Technology from Cochin University of Science and Technology (CUSAT), Cochin,

India, in 2014. His research interests are design and analysis of Streamlined airborne radomes, wireless communication, digital beamformers, structurally integrated radiating structures, and EM material characterization.

Dr. Raveendranath U. Nair is currently Senior Principal Scientist and Head in Centre for Electromagnetics (CEM), CSIR-National Aerospace Laboratories (CSIR-NAL), Bangalore, India. He received M.Sc. in Physics and Ph.D. in Physics (Microwave Electronics) from the School of Pure and Applied Physics, Mahatma Gandhi University, Kerala, India, in 1989 and 1997, respectively. He has authored/co-authored over 150 research publications including peer-reviewed journal papers, symposium papers, and technical reports. He has co-authored a chapter in a book *Sensors Update* published by Wiley-VCH, Germany, in 2000. The electromagnetic (EM) material characterization techniques developed for his doctoral work were included in the section *Perturbation Theory* in RF and Microwave Encyclopedia (vol. 4) published by John Wiley & Sons, USA, in 2005. He received the *CSIR-NAL Excellence in Research Award* (2007–2008) for his contributions to the EM design of variable thickness airborne radomes. He was selected for the *CSIR Leadership Development Programme (CSIR-LDP)* in 2008. He is also a Professor at the Academy of Scientific and Innovative Research (AcSIR), New Delhi.

Abbreviations and Symbols

Abbreviations

BSE	Boresight error
CTR	Constant thickness radome
EBSE	Elevation boresight error
EM	Electromagnetics
ETLM	Equivalent transmission line method
IPL	Inhomogeneous planar layer
PU	Polyurethane
SLL	Side lobe level
TE	Transverse electric
V_{AZ}	Azimuth difference error voltages
V_{EL}	Elevation difference error voltages
VTR	Variable thickness radome
Δ_{AZ}	Azimuth difference channel voltages
Δ_{EL}	Elevation difference channel voltages

Symbols

ε_r	Relative permittivity
$\tan \delta e$	Electric loss tangent
E_t	Electric field transmitted by slotted waveguide planar array antenna
H_t	Magnetic field transmitted by slotted waveguide planar array antenna
E_r	Electric field corresponding to received signal
H_r	Magnetic field corresponding to received signal
V_{port}	Port voltage
$\hat{n}$	Unit vector
C	Integral constant
J_r	Induced electric current
M_r	Induced magnetic current
Δ_{AZ}	Azimuth difference channel voltages

Δ_{EL}	Elevation difference channel voltages
ΔV_{AZ}	Azimuth difference error voltage
ΔV_{EL}	Elevation difference error voltage
$\sum$	Sum port voltage
T_{mn}	Complex transmission coefficient of the mnth radiating slot
A_{mn}	Aperture distribution function of the mnth radiating slot
k	Propagation constant
x_m	x-coordinate of the mnth radiating slot
y_n	y-coordinate of the mnth radiating slot
K	Monopulse antenna sensitivity constant

List of Figures

Abstract

The stringent electromagnetic (EM) performance requirements of Streamlined airborne radomes demand development of new radome wall designs such as graded dielectric inhomogeneous planar structure as proposed in this book. The analysis of the proposed radome-enclosed slotted waveguide antenna system is carried out from which it can be concluded that the radome exhibits good performance over X-band with center frequency 10 GHz and with power transmission >90%.

Chapter 1
Introduction

The design of airborne streamlined radomes is a daunting task as it has to encounter the constraints in fabrication and also provides sufficiently high performance (Burks 2007; Kozakoff 2010; Nair and Jha 2014). Controllable artificial dielectric structures play a vital role in the design of such radome wall structures (Nair and Jha 2007; Nair and Jha 2009). The efficacy of variable thickness radome (VTR) to provide superior power transmission capability makes it a suitable candidate than constant thickness radomes (CTR) (Nair and Jha 2009; Nair et al. 2014). Even with the advent of technology, fabrication of such streamlined airborne radomes with controllable dielectric parameters is an arduous job and may degrade the radome performance. To circumvent this limitation in fabrication and to enhance the performance parameters of the radome, a novel radome wall structure based on inhomogeneous planar layer (IPL) has been reported in this book.

The effect of designing a tangent-ogive radome wall using an inhomogeneous planar layer to provide graded variation of dielectric parameters has been already demonstrated by the author (Nair et al. 2015). The antenna considered in this work was a planar slotted waveguide array antenna with cosine distribution. This study is further extended in the present work for other aperture distributions as well, and the performance of the radome with the inclusion of a radome paint so as to retain its superior EM characteristics even in hoarse environmental conditions has also been evaluated. Moreover, a hemispherical radome with this novel wall configuration for ground-based applications has been analyzed in this work. The performance of the radome enclosing slotted waveguide planar array antenna is compared with the conventional optimized radome designs such as constant thickness radome (CTR) and variable thickness radome (VTR) designs. Further the computed radiation characteristics of the graded dielectric inhomogeneous radome enclosed antenna system at preselected orientation of antenna in both the elevation and azimuth plane are analyzed that shows minimal degradations (in terms of beam width, SLL, and flash lobes), which is desirable for radome applications.

© The Author(s) 2018
P. Mahima et al., *Electromagnetic Performance Analysis of Graded Dielectric Inhomogeneous Radomes*, SpringerBriefs in Applied Sciences and Technology, https://doi.org/10.1007/978-981-10-7832-3_1

Chapter 2
Radome Wall Configurations

The design of radome wall is a critical task in the manufacturing of airborne vehicles, which has to adhere to the electromagnetic boundary conditions and the availability of radome materials. In addition, the radome has to survive the effect of rain, erosion, lightning strikes, aerodynamic loads, and other destructive factors. The choice of type of radome wall construction will depend on the application requirements. A single thin skin may be suitable at low frequencies, but multiples of half-wave thickness may be essential to provide sufficient strength at the higher frequencies, where the sandwich types may be preferred as they provide high strength-to-weight ratio and wide bandwidths.

At various wavelengths, while evaluating the radome electrical performance it is necessary to consider the electrical properties of radome wall materials. These properties consist of mainly dielectric constant and loss tangent. There are different types of radome dielectric wall materials present.

Radome wall construction is basically one of the following designs: Solid monolithic, generally, made of resins incorporating reinforcements such as chopped glass fibers, sandwich design consisting of alternating high-density (high $\in_R$) and low-density (low $\in_R$) materials (Kozakoff 2010). We obtain the resulting relative dielectric constant of the mixture from the following equation:

$$\in_m = \frac{V_R \log \in_R + V_F \log \in_F}{V_R + V_F}$$

where

$\in_m$ relative dielectric constant of mixture
$\in_R$ relative dielectric constant of resin
$\in_F$ relative dielectric constant of reinforcement fibers
V_R volume of the resin
V_F volume of the reinforcement fibers

This equation assumes the mixture to be uniform and isotropic.

© The Author(s) 2018
P. Mahima et al., *Electromagnetic Performance Analysis of Graded Dielectric Inhomogeneous Radomes*, SpringerBriefs in Applied Sciences and Technology, https://doi.org/10.1007/978-981-10-7832-3_2

Monolithic: A monolithic wall (Fig. 2.1a) is made from a single material, either
                 thin wall or multiple one-half wavelength thick dielectrics, with less
                 reflection.

A-sandwich: A-sandwich (Fig. 2.1b) consists of two skins having higher
                 dielectric constant than that of core, and electrically very thin of
                 half a wavelength thickness. They are separated by a lower
                 dielectric constant core of such a thickness so as to yield substantial

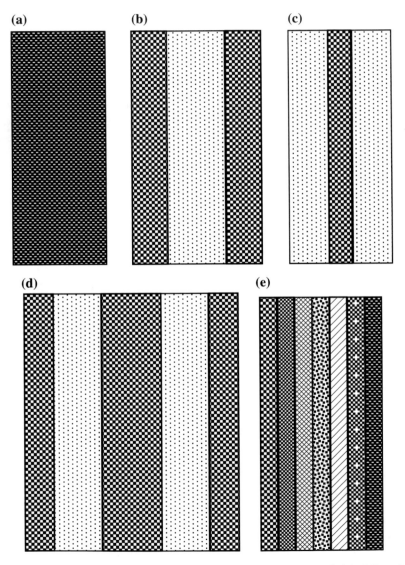

Fig. 2.1 Radome wall configurations: **a** Monolithic, **b** A-sandwich, **c** B-sandwich, **d** C-sandwich,
e Multilayered

cancelation of the skin reflections. In this construction, the strength-to-weight ratio and bandwidth are greater than in a solid-wall/monolithic radome.

B-sandwich: This wall configuration (Fig. 2.1c) consists of two skins having low dielectric constant and a thin solid core of high dielectric constant. B-sandwich is designed to provide equal transmission for perpendicular and parallel polarizations. Here the skins act as quarter-wave transformer.

C-sandwich: C-sandwich (Fig. 2.1d) consists of five layers, i.e., two A-sandwich placed back-to-back without spacing between them. This construction gives the best strength-to-weight ratio. Here transmission efficiency and bandwidth are greater than that of A-sandwich. The residual reflections from the individual A-sandwiches are further canceled. It is having two outer skins having equal dielectric constant and loss tangent, two cores having same dielectric constants and loss tangent and an inner skin of higher thickness, dielectric constant, and loss tangent.

Multilayered: The multilayered wall (Fig. 2.1e) consists of very thin skins, cores of different dielectric constants which can give best strength and wide-band transmission properties for small angles of incidence. This type of radome wall configuration gives the high structural reliability, good electrical performance over wider bandwidths and better resistance to thermal shock and rainfall.

Chapter 3
EM Characteristics of Planar Slab Model of Graded Dielectric Inhomogeneous Planar Structure

In this brief, the equivalent transmission line model is mainly used for the EM designing of radome wall structure. It is the model in which the radome wall is considered as a transmission line and different sections similar to different layers (Fig. 3.1).

Here the low impedance lines are connected in series corresponding to the different layers of the radome wall structure. The impedances of different layers are taken as Z_1, Z_2, ..., Z_n. By taking the multiplication of each matrix corresponding to each layer, a single matrix is found which represents the whole configuration of the radome wall. So, the entire radome wall configuration may be expressed as (Refer Fig. 4.1)

$$\begin{bmatrix} A & B \\ C & D \end{bmatrix} = \begin{bmatrix} A_1 & B_1 \\ C_1 & D_1 \end{bmatrix} \begin{bmatrix} A_2 & B_2 \\ C_2 & D_2 \end{bmatrix} \begin{bmatrix} A_3 & B_3 \\ C_3 & D_3 \end{bmatrix} \begin{bmatrix} A_4 & B_4 \\ C_4 & D_4 \end{bmatrix} \begin{bmatrix} A_5 & B_5 \\ C_5 & D_5 \end{bmatrix} \begin{bmatrix} A_6 & B_6 \\ C_6 & D_6 \end{bmatrix} \begin{bmatrix} A_7 & B_7 \\ C_7 & D_7 \end{bmatrix}$$

$$(3.1)$$

The A, B, C, D parameters of each layer depend on dielectric constant, loss tangent, thickness of the layer, incidence angle, and the polarization, etc.

The matrix representation of the layer 1 is given by (Cary 1983)

$$\begin{bmatrix} A_1 & B_1 \\ C_1 & D_1 \end{bmatrix} = \begin{bmatrix} \cos \phi_1 & j\frac{Z_1}{Z_0} \sin \phi_1 \\ j\frac{Z_0}{Z_1} \sin \phi_1 & \cos \phi_1 \end{bmatrix}$$

$$(3.2)$$

Similarly, for all the layers of radome wall configuration, the matrices are represented by

$$\begin{bmatrix} A_2 & B_2 \\ C_2 & D_2 \end{bmatrix} = \begin{bmatrix} \cos \phi_2 & j\frac{Z_2}{Z_0} \sin \phi_2 \\ j\frac{Z_0}{Z_2} \sin \phi_2 & \cos \phi_2 \end{bmatrix}$$

$$(3.3)$$

© The Author(s) 2018
P. Mahima et al., *Electromagnetic Performance Analysis of Graded Dielectric Inhomogeneous Radomes*, SpringerBriefs in Applied Sciences and Technology,
https://doi.org/10.1007/978-981-10-7832-3_3

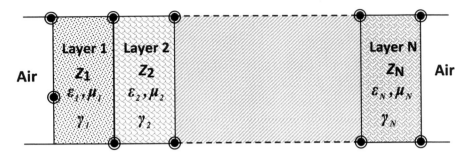

Fig. 3.1 Equivalent transmission line model of radome wall with N no. of layers

$$\begin{bmatrix} A_3 & B_3 \\ C_3 & D_3 \end{bmatrix} = \begin{bmatrix} \cos\phi_3 & j\frac{Z_3}{Z_0}\sin\phi_3 \\ j\frac{Z_0}{Z_3}\sin\phi_3 & \cos\phi_3 \end{bmatrix} \tag{3.4}$$

$$\begin{bmatrix} A_4 & B_4 \\ C_4 & D_4 \end{bmatrix} = \begin{bmatrix} \cos\phi_4 & j\frac{Z_4}{Z_0}\sin\phi_4 \\ j\frac{Z_0}{Z_4}\sin\phi_4 & \cos\phi_4 \end{bmatrix} \tag{3.5}$$

$$\begin{bmatrix} A_5 & B_5 \\ C_5 & D_5 \end{bmatrix} = \begin{bmatrix} \cos\phi_5 & j\frac{Z_5}{Z_0}\sin\phi_5 \\ j\frac{Z_0}{Z_5}\sin\phi_5 & \cos\phi_5 \end{bmatrix} \tag{3.6}$$

$$\begin{bmatrix} A_6 & B_6 \\ C_6 & D_6 \end{bmatrix} = \begin{bmatrix} \cos\phi_6 & j\frac{Z_6}{Z_0}\sin\phi_6 \\ j\frac{Z_0}{Z_6}\sin\phi_6 & \cos\phi_6 \end{bmatrix} \tag{3.7}$$

$$\begin{bmatrix} A_7 & B_7 \\ C_7 & D_7 \end{bmatrix} = \begin{bmatrix} \cos\phi_7 & j\frac{Z_7}{Z_0}\sin\phi_7 \\ j\frac{Z_0}{Z_7}\sin\phi_7 & \cos\phi_7 \end{bmatrix} \tag{3.8}$$

Here the EM performance parameters are considered only for perpendicular polarization. The Z/Z_0 for perpendicular polarization, parallel polarization, and electrical length ϕ corresponding to each dielectric layer is given, respectively, by

$$\frac{Z}{Z_0} = \frac{\cos\theta}{\sqrt{\varepsilon - \sin^2\theta}} \quad \text{for perpendicular} \tag{3.9}$$

$$\frac{Z}{Z_0} = \frac{\sqrt{\varepsilon - \sin^2\theta}}{\varepsilon\cos\theta} \quad \text{for parallel} \tag{3.10}$$

$$\phi = \frac{2\pi d\sqrt{\varepsilon - \sin^2\theta}}{\lambda} \tag{3.11}$$

where ε, d, and θ is the dielectric constant, thickness, and incidence angle, respectively.

Figure 3.2 shows the effective impedance for both perpendicular and parallel polarization of graded dielectric inhomogeneous structure with respect to antenna scan angle. For perpendicular polarization, the effective impedance is gradually decreasing according to the antenna scan angle. For parallel polarization, the effective impedance is gradually increasing with respect to antenna scan angle.

The power transmission coefficient is given by

$$P_{tr} = \left[\frac{4}{(A+B+C+D)^2} \right] \qquad (3.12)$$

The power reflection coefficient is given by

$$P_{rf} = \left[\frac{A+B-C-D}{A+B+C+D} \right]^2 \qquad (3.13)$$

The phase distortions are obtained by the *insertion phase delay* (IPD) of the radome wall. For a radome wall with seven layers, the insertion phase delay is given by

$$
\begin{aligned}
\text{IPD} = {} & -(\angle T_1 + \angle T_2 + \angle T_3 + \angle T_4 + \angle T_5 + \angle T_6 + \angle T_7) \\
& - \frac{2\pi}{\lambda}(d_1 \cos\theta_1 + d_2 \cos\theta_2 + d_3 \cos\theta_3 + d_4 \cos\theta_4 \\
& + d_5 \cos\theta_5 + d_6 \cos\theta_6 + d_7 \cos\theta_7)
\end{aligned}
\qquad (3.14)
$$

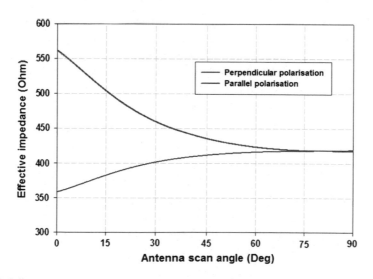

Fig. 3.2 Effective impedance of graded dielectric inhomogeneous planar structure for perpendicular and parallel polarization

Here $\angle T_1$, $\angle T_2$, $\angle T_3$, $\angle T_4$, $\angle T_5$, $\angle T_6$, $\angle T_7$ are the phase angles associated with the voltage transmission coefficients of the dielectric layers with thickness d_1, d_2, d_3, d_4, d_5, d_6, d_7, respectively. θ_1, θ_2, θ_3, θ_4, θ_5, θ_6, θ_7 are the corresponding incidence angles at each interface of the dielectric layers.

Figure 3.3 shows the power transmission, power reflection, and insertion phase delay of planar slab model for graded dielectric inhomogeneous planar structure with respect to frequency for different incidence angles. At normal incidence angle, the power transmission is high (>98%). For 45° angle of incidence, the power transmission is almost 96%. At the incidence angle of 60°, the power transmission is nearly 92% as shown in Fig. 3.3a. According to Fig. 3.3b, the power reflection is very less at normal incidence angle. For the incidence angle of 45°, the power reflection is approximately −15 dB and at 60° angle of incidence, the power reflection is almost −12 dB.

The insertion phase delay of a planar slab model for graded dielectric inhomogeneous planar structure (Fig. 3.3c) at angle of incidence 60° is observed to be the highest when compared to that at angle of incidence 0° and 45°. It indicates that at nosecone sector, maximum phase distortion is observed.

Power transmission, power reflection, and insertion phase delay for planar slab model of graded dielectric inhomogeneous planar structure with respect to frequency range of 1–18 GHz, for different incidence angle is shown in Fig. 3.4. At normal incidence, the power transmission is very high (>98%) and for 45° and 60° it is around 95%. So overall for the frequency range of (1–7 GHz) the power transmission for the entire range of incidence angle is found to be more than 95%. At X-band (8–12 GHz) for 0° incidence angle, the power transmission is more than 97%. For an incidence angle of 45°, the power transmission is almost 94% for x-band (8–12 GHz) and with the same frequency band the power transmission at 60° angle of incidence is nearly from 90 to 94%. For Ku band, the power transmission at normal incidence is high (>96%) and for the same frequency band for the incidence angle of 45°, the power transmission is more than 92%.

According to Fig. 3.4b, the power reflection at normal incidence for the frequency range of (1–18) GHz is very less (<3%) and for the same frequency range the power reflection at 45° and 60° incidence angle is less (10%). According to Fig. 3.3b, the power reflection and the insertion phase delay at normal incidence for the frequency range of (1–18) GHz is very less (<3% and <15°) and for the same frequency range these performance parameters at 45° and 60° incidence angle shows comparatively low value (<10% and <30°).

In this brief, a tangent ogive airborne radome is proposed based on graded inhomogeneous dielectric structure which consists of seven layers ($N = 7$ layers), with the middle layer having the maximum dielectric constant which decreases in a graded or step-wise manner on both sides of the middle layer. The theory of small reflections is used for the design of the symmetrical structure. The analysis for the reflection coefficient at each layer can be obtained by

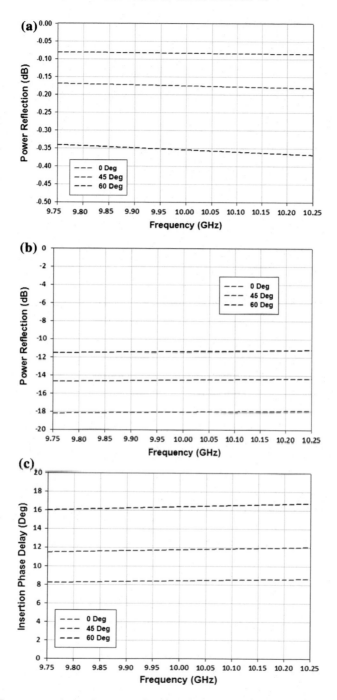

Fig. 3.3 **a** Power transmission, **b** power reflection, **c** insertion phase delay of planar slab model of graded dielectric inhomogeneous planar structure

Fig. 3.4 **a** Power transmission, **b** power reflection, **c** insertion phase delay of planar slab model of graded dielectric inhomogeneous planar structure for frequency range of (1–18) GHz

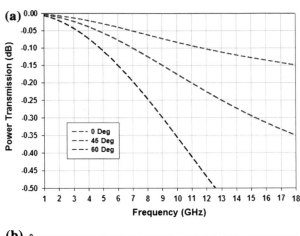

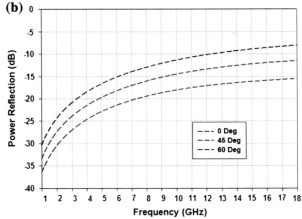

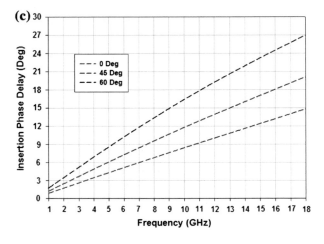

$$\Gamma_0 = \frac{\eta_1 - \eta_0}{\eta_1 + \eta_0} \tag{3.15}$$

$$\Gamma_1 = \frac{\eta_2 - \eta_1}{\eta_2 + \eta_1} \tag{3.16}$$

$$\vdots$$

$$\Gamma_N = \frac{\eta_{N+1} - \eta_N}{\eta_{N+1} + \eta_N} \tag{3.17}$$

where η is the intrinsic impedance of different layers and is given as

$$\eta_0 = \frac{\varepsilon_0}{\mu_0} \tag{3.18}$$

$$\eta_1 = \eta_0 \eta_2 = \sqrt{\frac{\varepsilon_0 \varepsilon_2}{\mu_0 \mu_2}} \tag{3.19}$$

$$\eta_2 = \eta_0 \eta_3 = \sqrt{\frac{\varepsilon_0 \varepsilon_3}{\mu_0 \mu_3}} \tag{3.20}$$

$$\vdots$$

$$\eta_N = \eta_0 \eta_{N+1} = \sqrt{\frac{\varepsilon_0 \varepsilon_{N+1}}{\mu_0 \mu_{N+1}}} \tag{3.21}$$

where ε is the dielectric constant of different layers, (as shown in Fig. 3.1) and μ is the permeability of different layers. Here the ε_0 and μ_0 are the dielectric constant and permeability of air, which are constant ($\varepsilon_0 = 8.85 \times 10^{-12}$ F/m and $\mu_0 = 4\pi \times 10^{-7}$ H/m).

Chapter 4
EM Design Aspects of Graded IPL

The proposed radome wall is designed in such a way that it consists of seven dielectric layers cascaded together (as shown in Fig. 4.1) such that the center layer (glass epoxy) is having maximum dielectric constant and electric loss tangent which then varies in a graded or step-wise manner on either sides of the center layer (except that of Layer 1 and Layer 7).

The different materials used for materializing the graded dielectric wall structure and their material parameters are tabulated in Table 4.1.

4.1 Tangent Ogive Airborne Radome: EM Design Aspects

A tangent ogive-shaped radome with a height of 1000 mm and base diameter 500 mm is considered. It is assumed to be enclosing a monopulse planar array antenna (perpendicularly polarized) with circular aperture of diameter 300 mm located at a height of 300 mm from the radome base. The antenna is operating at a frequency of 10 GHz with a bandwidth of 500 MHz. The spacing between the slots is taken as 15 mm and aperture distribution of the antenna as cosine in nature. Triangular lattice geometry of the array elements is taken into consideration in this study. The azimuth and elevation scan ranges of the antenna are ±70°.

The most critical parameters to be evaluated for a radome are power transmission efficiency and boresight error. Maximum power transmission and minimum BSE can be achieved by creating a variable thickness radome (VTR), wherein the thickness of the radome wall is optimized to get superior EM performance characteristics. In this section, the design of the conventional VTR radomes has been discussed in detail. Monolithic half-wave VTR is made of glass epoxy (dielectric constant, $\varepsilon_r = 4.0$, electric loss tangent, $\tan \delta_e = 0.015$), and the thickness is varying gradually from radome nose-tip to base (8.3–7.4 mm). A-sandwich VTR considered here has two constant thickness skin layers made of glass epoxy (thickness = 0.75 mm). The thickness of core layer, made of PU Foam ($\varepsilon_r = 1.15$,

© The Author(s) 2018
P. Mahima et al., *Electromagnetic Performance Analysis of Graded Dielectric Inhomogeneous Radomes*, SpringerBriefs in Applied Sciences and Technology, https://doi.org/10.1007/978-981-10-7832-3_4

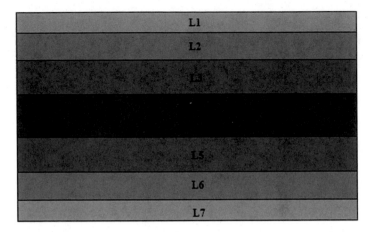

Fig. 4.1 Graded dielectric inhomogeneous radome: variation of dielectric parameters across the wall

Table 4.1 Different layers of the graded dielectric structure

Layer	Material	Dielectric constant (ε_r)	Electric loss tangent (tan δ)	Thickness (t)
L_1, L_7	Quartz-based lamination	1.04	0.0008	0.4
L_2, L_6	Quartz honeycomb	1.06	0.0003	3.1
L_3, L_5	PU Foam	1.15	0.0020	0.5
L_4	Glass epoxy	4	0.015	0.3

electric loss tangent, tan δ_e = 0.002), is varying from 9.8 mm at nose to 8.5 mm at base. In the case of C-sandwich, the skin layers and middle layer are all made of glass epoxy (thickness of outer skin layers = 0.75 mm and thickness of middle skin = 1.5 mm). The thickness of both core layers, fabricated of PU Foam, is varying gradually from radome nose-tip to base (4.9–4.6 mm) (Fig. 4.2).

Fig. 4.2 Tangent–ogive-shaped airborne radome with planar array antenna

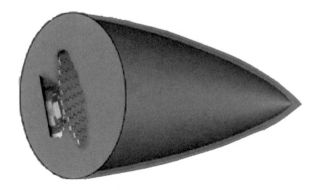

Fig. 4.3 Hemispherical-
shaped radome with planar
array antenna

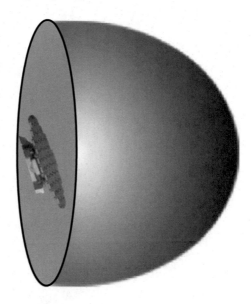

4.2   Hemispherical Airborne Radome: EM Design Aspects

A hemispherical radome having dimensions (height = 500 mm; base diameter = 1000 mm) enclosing a monopulse planar array antenna with circular aperture (diameter = 250 mm) located at a height of 150 mm from the radome base is considered here (Fig. 4.3). The antenna is operating at a frequency of 10 GHz with a bandwidth of 500 MHz and is perpendicularly polarized. The aperture distribution of the antenna is cosine in nature. The spacing between the slots of the monopulse planar array antenna is taken as 15 mm. The design of CTR profiles for the conventional wall configurations for hemispherical radome is detailed in this section. Monolithic half-wave radome is composed of glass epoxy with a thickness of 4.97 mm. A-sandwich is having two skin layers made of glass epoxy of thickness 0.75 mm each and one core layer made of PU Foam of thickness 3.239 mm. In C-sandwich, skin layers are made of glass epoxy (thickness of outer skin layers = 0.75 mm, thickness of middle skin = 1.5 mm) and two core layers fabricated of PU Foam of thickness 3.239 mm each.

Chapter 5
EM Performance Analysis of Radomes

The study of antenna-radome interaction is performed by using 3D ray-tracing procedure based on geometrical optics along with aperture integration method (Kozakoff 2010; Nair et al. 2014). The procedure for EM performance analysis is as follows: (i) Determination of the coordinates of the antenna elements (ii) Transformation of the antenna coordinates to radome coordinates. (iii) Assigning of excitation to the antenna elements (iv) Computation of the transmission coefficient of the radome wall (v) Computation of far-field radiation pattern and radome performance parameters w.r. t antenna scan angle. The receiving mode approach is utilized to compute the performance parameters of the antenna such as co-pol, cross-pol, and boresight error. The efficacy of the transmitting mode approach is exploited to compute the far-field radiation pattern of the antenna-radome system which points out the degradation of the antenna radiation pattern.

In the receiving mode approach (backward ray-trace method), an aperture surface in similar to the antenna aperture is assumed external to radome. The parallel rays emanating from such an external aperture are projected to antenna's aperture plane. While computing the radiation pattern and the performance parameters, to include the effect of gimbal system, the antenna coordinates need to be transformed w.r.t radome coordinates. Here the gimbal system considered is free from rotational offsets. The angle of incidence corresponding to each ray from the slots of the antenna to the inner surface of radome is determined to evaluate the co-pol and cross-pol voltage transmission coefficients based on the dielectric parameters of the radome wall. By integrating the fields on the antenna aperture, the receiving port voltages are computed.

The sum and difference channel port voltages are computed based on Lorentz reciprocity theorem (Huddleston et al. 1978). The electric and magnetic fields radiated by the slots of the antenna are represented by E_t and H_t. Let E_r and H_r be the fields for the plane wave illumination resulting from the return echo signal. Then the port voltage is represented by

P. Mahima et al., *Electromagnetic Performance Analysis of Graded Dielectric Inhomogeneous Radomes*, SpringerBriefs in Applied Sciences and Technology, https://doi.org/10.1007/978-981-10-7832-3_5

$$V_{\text{port}} \propto \left[\int\limits_s \left(\vec{E}_t \times \vec{H}_r \right) \cdot \hat{n} - \left(\vec{E}_r \times \vec{H}_t \right) \cdot \hat{n} \right] \mathrm{d}s \qquad (5.1)$$

Then

$$V_{\text{port}} = C \left[\int\limits_s \left(\vec{E}_t \times \vec{H}_r \right) \cdot \hat{n} - \left(\vec{E}_r \times \vec{H}_t \right) \cdot \hat{n} \right] \mathrm{d}s \qquad (5.2)$$

where the term C is inversely proportional to the current coefficient corresponding to fields of return echo signal.

The closed surface S encloses all the excitation sources (radiating slots of antenna and fields induced on inner surface of radome) responsible for the pair of fields (E_t, H_t) and (E_r, H_r). Equation (5.2) is further modified by vector operation along the closed surface S and the replacement of return plane wave excitation fields by induced currents J_r and M_r based on equivalence principle. Then resultant port voltage is obtained by

$$V_{\text{port}} = C \int\limits_s \left[\vec{E}_t \cdot \vec{J}_r - \vec{H}_t \cdot \vec{M}_r \right] \mathrm{d}s \qquad (5.3)$$

Equation (5.3) is used for the determination of the monopulse sum channel voltage ($\sum$), and difference channel voltages (Δ_{AZ} and Δ_{EL}). Co-pol and cross-pol power transmission parameters are calculated using sum channel voltage. From the difference channel voltages and sum channel voltage, the monopulse azimuth and elevation plane error voltages (ΔV_{AZ} and ΔV_{EL}) are obtained. The boresight errors for azimuth and elevation scans are given by (Huddleston et al. 1978; Siwiak et al. 1979)

$$\text{BSE}_{\text{AZ}} = \frac{\Delta V_{\text{AZ}}}{K_{\text{AZ}}} \qquad (5.4)$$

$$\text{BSE}_{\text{EL}} = \frac{\Delta V_{\text{EL}}}{K_{\text{EL}}} \qquad (5.5)$$

Here K represents the monopulse antenna sensitivity constants, which depends on the difference channel port voltages and azimuth/elevation offsets. The unit of K is volts per mrad.

In the transmitting mode, the focus is on the EM wave from the antenna aperture that propagates through the radome wall in a direction normal to the antenna surface, to an equivalent aperture outside of the radome. The point of intersection of the ray with the curved inner surface of the radome wall is approximated as a plane slab. The far-field pattern of the antenna alone is obtained as (Balanis 2005)

$$E_t = \sum_{m=1}^{M} \sum_{n=1}^{N} A_{mn} e^{-jk(x_m \sin\theta \cos\phi + y_n \sin\theta \sin\phi)} \tag{5.6}$$

where A_{mn} is the aperture distribution function of the mn-th radiating slot (cosine distribution in this case). The x and y coordinates of the mn-th radiating slot are represented by x_m and y_n.

While computing the far-field of the antenna-radome system, to include the effect of radome wall a term called transmission coefficient is taken into account. The incident field at each ray-radome intercept point is modified by the amplitude and phase of its associated radome transmission coefficients computed based on a "local flat panel" approximation of the radome wall. The far-field radiation pattern of the antenna-radome is computed by integration of the tangential fields on the external equivalent aperture. The far-field radiation pattern of antenna-radome system is given by

$$E_t = \sum_{m=1}^{M} \sum_{n=1}^{N} T_{mn} A_{mn} e^{-jk(x_m \sin\theta \cos\phi + y_n \sin\theta \sin\phi)} \tag{5.7}$$

where T_{mn} is the complex valued transmission coefficient associated with the intercept of the ray on the radome wall corresponding to mn-th radiating slot on the antenna surface.

Schematic of constant thickness radome (CTR) and variable thickness Radome (VTR) structures are shown in Fig. 5.1. In CTR designs, the radome wall thickness is kept constant from the radome nose-tip to radome base. In contrast to this, the wall thickness is gradually decreasing from radome nose-tip to radome base (nose thick-base thin).

A comparative study of constant thickness and variable thickness radome designs is carried out based on the EM performance parameters. VTR design is superior in performance compared to CTR design, which makes it a better choice

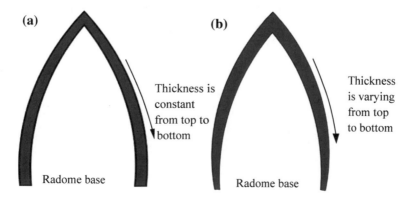

(a)

(b)

Thickness is constant from top to bottom

Thickness is varying from top to bottom

Radome base

Radome base

Fig. 5.1 Schematic of radome design: **a** CTR design; **b** VTR design

for airborne radome applications, but the fabrication of such structures with controlled electrical parameters is a strenuous task.

5.1 Comparative Study on the EM Performance Analysis of Graded Dielectric Radome by Full-Wave Method and ETL Method

The EM design of the radome wall based on a single method is not reliable as it may give inaccurate performance characteristics. To ascertain the authenticity, we validate the obtained results against those found by other methods. Here, besides using equivalent transmission line method, FDTD-based simulation is carried out in Microwave CST Studio Suite to ensure the accuracy of the computed characteristics. Both parallel polarization and perpendicular polarization are considered for simulation to trace the behavior of the structure w.r.t polarization. The structure is analyzed for various angle of incidence of the impinging ray by both methods and the performance curves are compared and shown in Figs. 5.2 and 5.3.

The detailed analysis of power reflection characteristics of the graded dielectric radome wall using equivalent transmission line method and FDTD-based simulation shows that they are in good agreement with each other, whereas its power transmission characteristics exhibit slight variations.

The power transmission curve for the graded IPL dielectric structure at an angle of incidence 0° is shown in Fig. 5.2a. It shows high transmission, over 98% which is desirable for radome applications. Figure 5.2b, c presents the transmission curves for the mid-section and nose-cone sector of the radome, i.e., at incidence angle 45° and 60°, respectively. The nose-cone sector plays a vital role in the designing of radome as it contributes maximum to BSE. Observation of the characteristics computed by both methods clearly depict that the power transmitted through the nose-cone sector is more than 90%.

The reflection characteristics of the novel structure are shown in Fig. 5.3 for normal incidence and for different oblique angle of incidence. The analysis implies that the reflection is minimum when angle of incidence is 0°. The reflection at incidence angle 45° and 60° is comparatively higher than that at normal incidence, but it is within tolerable limits.

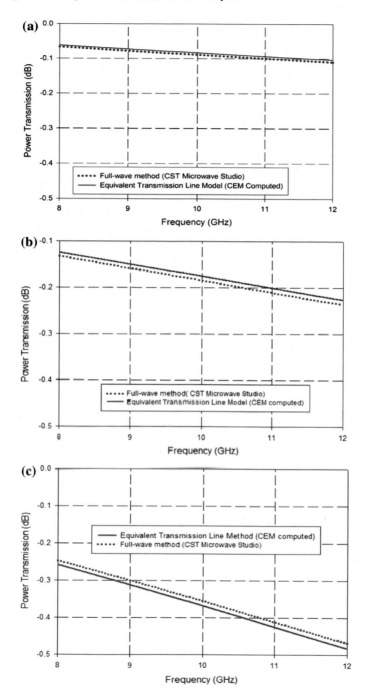

Fig. 5.2 Power transmission at an angle of incidence of **a** 0°, **b** 45°, **c** 60° for a planar slab model of graded dielectric inhomogeneous structure

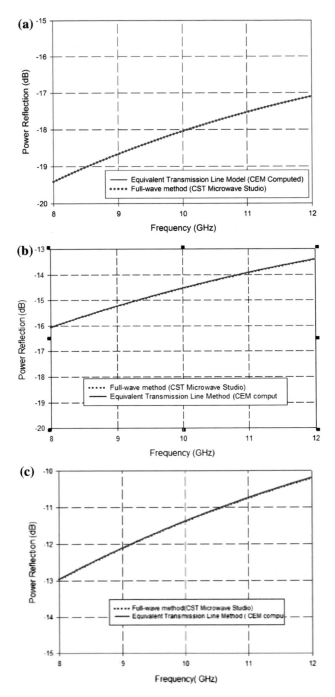

Fig. 5.3 Power reflection at an angle of incidence of **a** 0°, **b** 45°, **c** 60° for a planar slab model of graded dielectric inhomogeneous structure

5.2   Comparative Study on the EM Performance Analysis of Graded Dielectric Radome with Constant Thickness Radome (CTR) Designs

The EM performance characteristics of graded dielectric IPL radome, monolithic half-wave, A-sandwich, and conventional C-sandwich radome of identical thickness are shown in the performance curves given below. Here the EM performance parameters are analyzed for perpendicular (TE) polarization.

The co-pol power transmissions for different frequencies are given in Fig. 5.4. The power transmission of monolithic half-wave is less compared to graded dielectric inhomogeneous radome for all the frequencies. At 10 and 9.75 GHz graded dielectric inhomogeneous radome is having higher power transmission compared to C-sandwich in the nose-cone region (0°–10°) but from (10°–45°), the power transmission efficiency of C-sandwich is slightly better while from 45° onwards graded dielectric inhomogeneous radome shows better power transmission

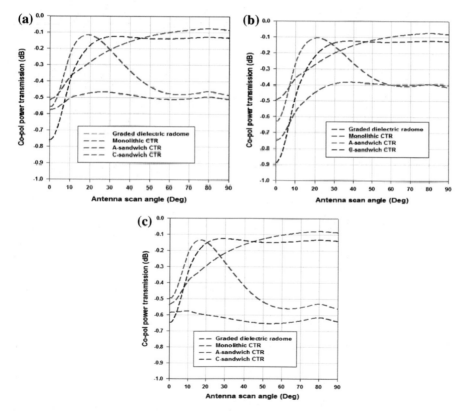

Fig. 5.4 Co-pol power transmission characteristics of streamlined nose-cone radome based on monolithic, A-sandwich, C-sandwich, and graded dielectric inhomogeneous radome design at **a** center frequency 10 GHz, **b** lower frequency 9.75 GHz, **c** higher frequency 10.25 GHz

characteristics. Compared to graded dielectric inhomogeneous radome, A-sandwich has superior power transmission efficiency in the antenna scan range of (5°–30°) for both 10 and 9.75 GHz, beyond 30°, power transmission efficiency of A-sandwich decreases rapidly as compared to that of graded dielectric radome.

Graded dielectric inhomogeneous radome is having higher power transmission compared to C-sandwich in the antenna scan range of 0°–5° but from 5° to 45°, the power transmission efficiency of C-sandwich is slightly better while from 45° onwards graded dielectric inhomogeneous radome shows better power transmission characteristics at 10.25 GHz. Compared to graded dielectric radome, A-sandwich has superior power transmission efficiency in nose-cone region (0°–20°). Beyond 20°, power transmission efficiency of A-sandwich decreases drastically as compared to that of graded dielectric inhomogeneous radome.

Figure 5.5 shows the X-pol power transmission characteristics of streamlined radome at frequencies 10, 9.75, and 10.25 GHz. The X-pol power transmission of

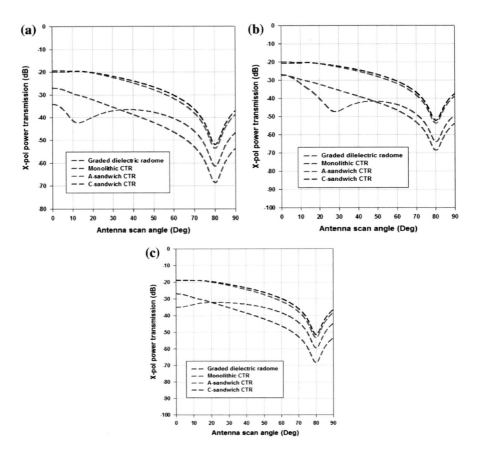

Fig. 5.5 X-pol power transmission characteristics of streamlined nose-cone radome based on monolithic, A-sandwich, C-sandwich, and graded dielectric inhomogeneous radome design at **a** center frequency 10 GHz, **b** lower frequency 9.75 GHz, **c** higher frequency 10.25 GHz

graded dielectric inhomogeneous radome is well below −30 dB throughout the antenna scan range, and it is superior to that of monolithic half-wave design for all the frequencies. For A-sandwich and C-sandwich design, the X-pol power transmission is much higher compared to that of graded dielectric radome which is adverse for airborne radar systems.

According to Fig. 5.6, the azimuth BSE of graded dielectric inhomogeneous radome design is slightly more as compared to that of monolithic half-wave in the nose-cone region (0°–15°) at 10 and 10.25 GHz and beyond 15°, graded dielectric radome shows better BSE characteristic compared to all other designs. However, the BSE of graded dielectric radome is within 3 mrad which is acceptable for airborne radome applications. At 9.75 GHz for the antenna scan range of (5°–20°) monolithic half-wave is having less boresight error compared to all other designs

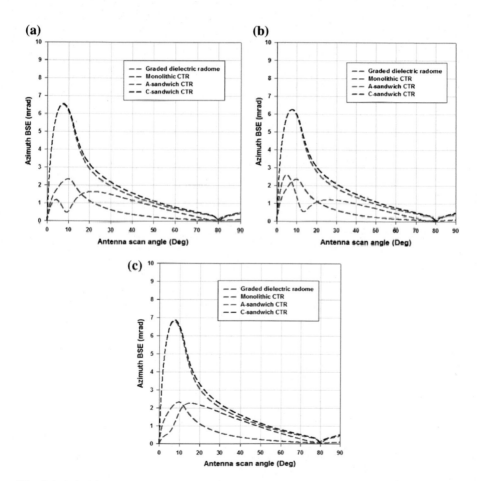

Fig. 5.6 Azimuth boresight error characteristics of streamlined nose-cone radome based on monolithic, A-sandwich, C-sandwich, and graded dielectric inhomogeneous radome design at **a** center frequency 10 GHz, **b** lower frequency 9.75 GHz, **c** higher frequency 10.25 GHz

but for all other angles graded dielectric radome is superior. For all the frequencies, A-sandwich and C-sandwich show very high BSE in the critical nose-cone sector (antenna scan range 0°–15°).

The performance curves are given in Fig. 5.7 and show the elevation BSE characteristics at different frequencies. The elevation BSE characteristics are almost similar to azimuth BSE characteristics. Graded dielectric radome is having the lowest EBSE compared to all other designs from the antenna scan angle of 10° onwards.

In the nose-cone region (0°–10°), monolithic half-wave is having less EBSE than that of graded IPL design for all the frequencies. Beyond 20°, the EBSE of monolithic half-wave is slightly higher compared to all other designs. In all the three cases, the EBSE of A-sandwich and C-sandwich increases drastically in the critical nose-cone sector (antenna scan range 0°–15°).

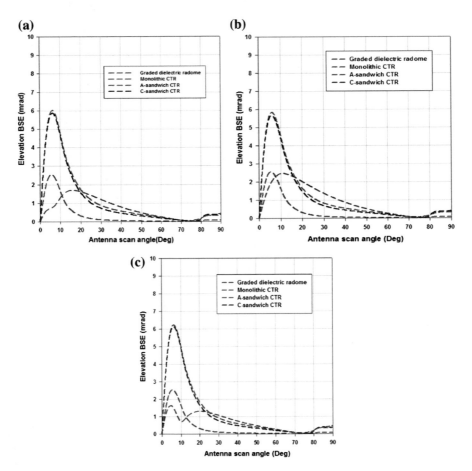

Fig. 5.7 Elevation boresight error characteristics of streamlined nose-cone radome based on monolithic, A-sandwich, C-sandwich, and graded dielectric inhomogeneous radome design at **a** center frequency 10 GHz, **b** lower frequency 9.75 GHz, **c** higher frequency 10.25 GHz

5.3   Comparative Study on the EM Performance Analysis of Graded Dielectric Radome with Variable Thickness Radome (VTR) Designs

Co-pol power transmission, X-pol power transmission, azimuth BSE, and elevation BSE characteristics of VTR design based on monolithic half-wave, graded dielectric, A-sandwich, and C-sandwich wall configurations are given from Figs. 5.8, 5.9, 5.10, and 5.11.

The co-pol power transmission (shown in Fig. 5.8) of monolithic half-wave is less compared to graded dielectric inhomogeneous radome at all the frequencies.

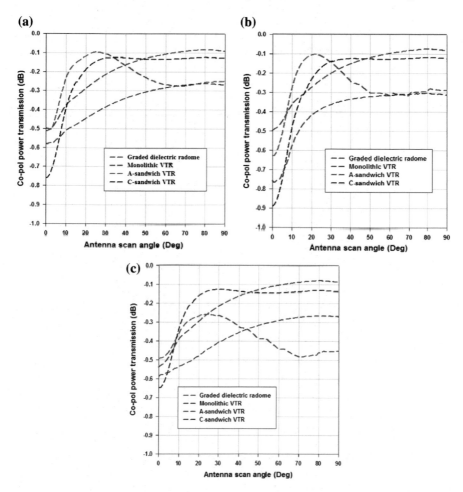

Fig. 5.8 Co-pol power transmission characteristics of streamlined nose-cone radome based on monolithic, A-sandwich, C-sandwich, and graded dielectric inhomogeneous variable thickness radome design at **a** center frequency 10 GHz, **b** lower frequency 9.75 GHz, **c** higher frequency 10.25 GHz

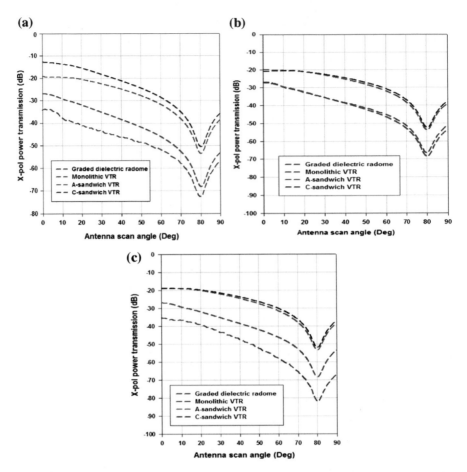

Fig. 5.9 X-pol power transmission characteristics of streamlined nose-cone radome based on monolithic, A-sandwich, C-sandwich, and graded dielectric inhomogeneous radome design at **a** center frequency 10 GHz, **b** lower frequency 9.75 GHz, **c** higher frequency 10.25 GHz

At 10 GHz, A-sandwich is having better power transmission characteristics up to antenna scan angle 40° but beyond that power transmission efficiency decreases rapidly. In the nose-cone region (0°–10°), power transmission of C-sandwich is less than that of graded dielectric inhomogeneous radome. In the antenna scan range of 10°–45°, C-sandwich shows slightly better power transmission. However, from 45° onwards graded dielectric radome has higher power transmission efficiency.

The curves at 9.75 GHz shows that A-sandwich is having higher power transmission in the antenna scan angle 5°–30° as compared to graded dielectric radome. Beyond 30°, power transmission efficiency of A-sandwich decreases. In the nose-cone region (0°–15°), graded dielectric inhomogeneous radome shows superior power transmission efficiency than C-sandwich. In the antenna scan range of 15°–45°, C-sandwich has slightly higher power transmission but from 45° onwards graded dielectric inhomogeneous radome gives better power transmission.

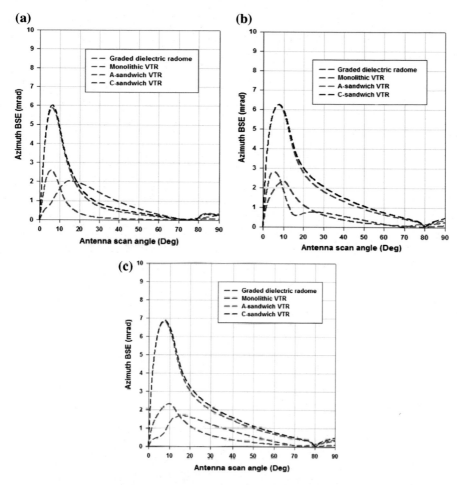

Fig. 5.10 Azimuth boresight error characteristics of streamlined nose-cone radome based on monolithic, A-sandwich, C-sandwich, and graded dielectric inhomogeneous radome design at **a** center frequency 10 GHz, **b** lower frequency 9.75 GHz, **c** higher frequency 10.25 GHz

For the higher frequency 10.25 GHz, graded dielectric inhomogeneous radome shows superior power transmission efficiency as compared to A-sandwich except in the antenna scan range 0°–20°. C-sandwich gives better power transmission than graded dielectric inhomogeneous radome from 5° to 45°, whereas for all other angles graded dielectric inhomogeneous radome is superior.

In case of lower frequency 9.75 GHz, graded dielectric radome outperforms all other designs in terms of X-pol transmission. It is even slightly better compared to monolithic half-wave from the antenna scan angle of 45° onwards.

According to the performance curves shown in Fig. 5.8, azimuth BSE of graded dielectric inhomogeneous radome is well within 3 mrad at all the frequencies which are acceptable, but for A-sandwich and C-sandwich designs azimuth BSE is above

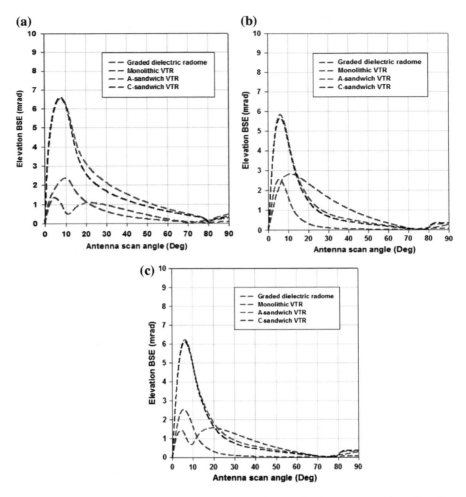

Fig. 5.11 Elevation boresight error characteristics of streamlined nose-cone radome based on monolithic, A-sandwich, C-sandwich, and graded dielectric inhomogeneous radome design at **a** center frequency 10 GHz, **b** lower frequency 9.75 GHz, **c** higher frequency 10.25 GHz

3 mrad especially in the critical nose-cone sector (0°–20°) which is undesirable for airborne radomes. The azimuth BSE of graded dielectric inhomogeneous radome is more as compared to monolithic half-wave at antenna scan angles (0°–15°) at 10 and 10.25 GHz and at (10°–25°) at 9.75 GHz.

The computed elevation BSE shows that for all the frequencies it is above 3 mrad in the nose-cone region (0°–15°) for A-sandwich and C-sandwich designs, which produces adverse effects on airborne applications. The elevation BSE of graded dielectric inhomogeneous radome is less (below 1 mrad) compared to all other designs at all frequencies except in the nose-cone region (0°–15°) where it is more (but less than 3 mrad) than that of monolithic half-wave design.

5.4 Comparative Study on the EM Performance Analysis of Graded Dielectric Inhomogeneous Structure for Hemispherical Radome

The EM performance of an antenna-radome system when the radome is hemispherical in shape is discussed in detail in this section. The different performance parameters such as co-pol power transmission, cross-pol power transmission, elevation boresight error, and azimuth boresight error are computed for different radome wall configurations and are compared with the proposed graded dielectric inhomogeneous radome wall. Figures 5.12 and 5.13 show the comparison graphs of these performance parameters for the conventional radome wall configuration and for the graded structure. It is clearly evident from Fig. 5.12a that the graded dielectric

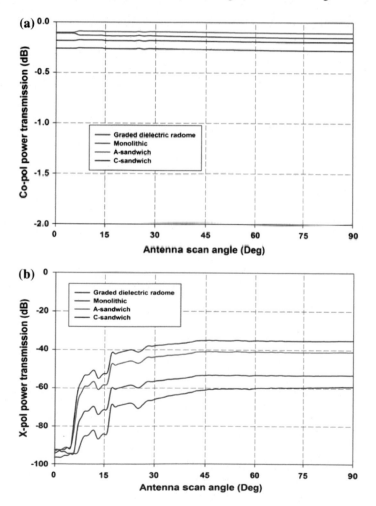

Fig. 5.12 Power transmission characteristics of graded dielectric radome with hemispherical shape **a** co-pol power transmission, **b** X-pol power transmission

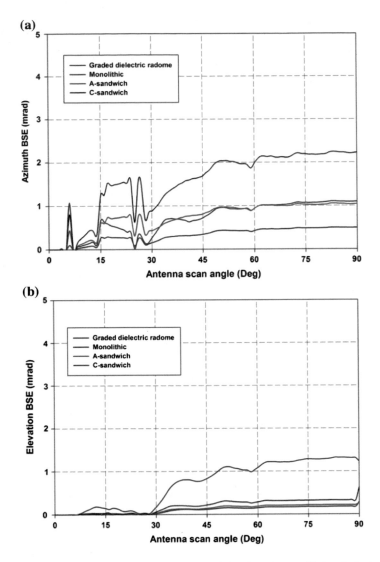

Fig. 5.13 Boresight error characteristics of graded dielectric radome with hemispherical shape **a** Azimuth BSE, **b** elevation BSE

inhomogeneous radome provides more than 95% power transmission, and it is almost constant over the whole scan range attributing to the negligible change in angle of incidence in comparison to normal incidence. But over the entire scan range, the A-sandwich wall exhibits supreme power transmission. The cross-pol power transmitted is less for the graded dielectric radome with hemispherical shape which is less than −50 dB that indicates the radome is performing better over the whole scan range (see Fig. 5.12b). The major advantage of using a hemispherical-shaped radome is its ability to exhibit low boresight error characteristics that is evident from the boresight graphs shown in Fig. 5.13a, b. The BSE in both elevation and azimuth planes is within 3 mrad. While comparing the boresight error produced by the hemispherical radome in antenna pattern for different wall configurations, it can be inferred that the C-sandwich structure produces maximum error in comparison to other walls and in the case of EBSE, monolithic shows high BSE. Even though it tends to be high, still it is in affordable limits.

5.5 EM Performance Analysis of Graded Dielectric Streamlined Radome for Different Distributions

The EM performance characteristics of graded dielectric inhomogeneous streamlined radome enclosed antenna with different aperture distributions are given below. In Fig. 5.14, the co-pol power transmission characteristics of graded dielectric inhomogeneous radome with different distributions at 10 GHz are shown. For all the distributions, the co-pol power transmission is almost similar at the scan range of 0°–30° but beyond the scan range of 30° the co-pol power transmission is exactly same for all the distributions. Figure 5.15 shows the X-pol power transmission for different distributions. For the entire scan range of 0°–90°, the X-pol power transmission for all distributions is almost similar.

Azimuth boresight error for different distributions is shown in Fig. 5.16. For Taylor distribution, the azimuth boresight error is completely 0 mrad for the entire scan range. The azimuth boresight error is almost constant at 0.9 mrad for uniform

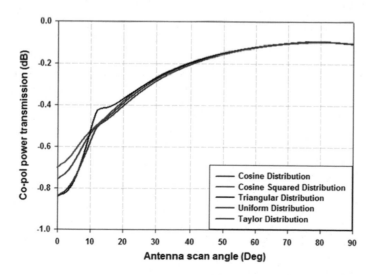

Fig. 5.14 Co-pol power transmission characteristics of graded dielectric streamlined radome with different aperture distribution (cosine, cosine squared, triangular, uniform and Taylor distribution)

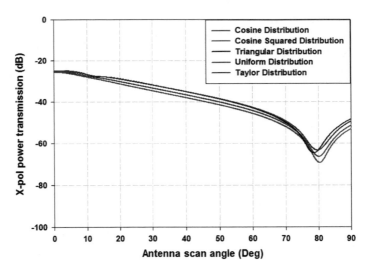

Fig. 5.15 X-pol power transmission characteristic of graded dielectric streamlined radome with different aperture distribution (cosine, cosine squared, triangular, uniform and Taylor distribution)

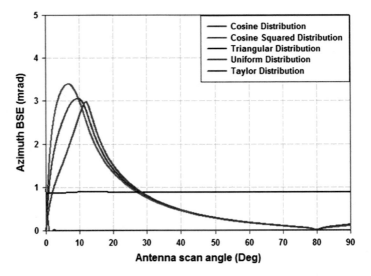

Fig. 5.16 Azimuth boresight error characteristic of graded dielectric streamlined radome with different aperture distribution (cosine, cosine squared, triangular, uniform and Taylor distribution)

distribution. For cosine, cosine-squared, and triangular distributions, the azimuth boresight error is within 3 mrad for the scan range of 0°–25°, but beyond the scan range of 25° the boresight error of the different distributions is exactly identical and less than 1 mrad.

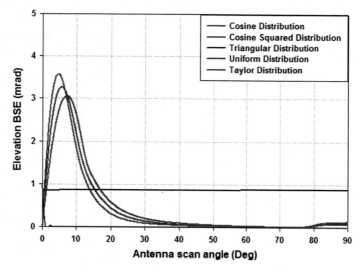

Fig. 5.17 Elevation boresight error characteristic of graded dielectric streamlined radome with different aperture distribution (cosine, cosine squared, triangular, uniform and Taylor distribution)

Figure 5.17 depicts the elevation boresight error for different distributions. The elevation boresight error is completely 0 mrad for Taylor distribution in the entire scan range. The elevation boresight error is almost constant at 0.9 mrad for uniform distribution. For cosine, cosine-squared, and triangular distributions, the elevation boresight error is within 3.5 mrad, but beyond the scan range of 25° the boresight error is 0 mrad for the three distributions.

5.6   EM Performance Characteristics of Graded Dielectric Inhomogeneous Streamlined Radome in the Entire Antenna Scan Region

In the previous section, EM performance characteristics of bare radome structures are compared. However, in actual operating conditions, the outer surface of the graded dielectric streamlined radome has to be protected from erosion and static charge effects due to aerodynamic drag. So the effect of a typical anti-static and anti-erosion radome paint ($\varepsilon_r = 3.46$; tan $\delta e = 0.068$; and thickness 0.2 mm) coated on the outer surface of the radome is considered here. Hence, the radome wall configuration has eight layers.

Co-pol power transmission, X-pol power transmission, and boresight error characteristics of the graded dielectric radome in the forward hemisphere are shown

(a)   **(b)**

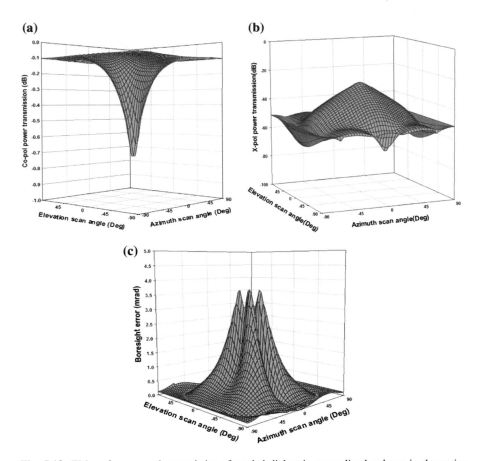

Fig. 5.18 EM performance characteristics of graded dielectric streamlined radome in the entire scan region at 10 GHz. **a** Co-pol power transmission, **b** X-pol power transmission, **c** boresight error

in Fig. 5.18. Co-pol power transmission efficiency is excellent in the entire forward hemisphere (>85%). The cross-pol transmission is well below −30 dB and boresight error is below 3 mrad, which are desirable for airborne application.

Figure 5.19 shows the comparison of radiation characteristics of *antenna alone* and *antenna-radome system* with the seeker antenna orientation at Az = 0°, El = 0° pointing at the nose-tip. Radiation characteristics of the antenna-radome system show no flash lobes and sidelobe-level degradations are also not austere, only the second and third sidelobes show minor variations and the first null is also not distinct. Thus, graded dielectric radome is a better choice for airborne radome applications. In Figs. 5.20 and 5.21, a comparison of radiation characteristics of *antenna alone* and *antenna-radome system* with gimbal orientation at four different

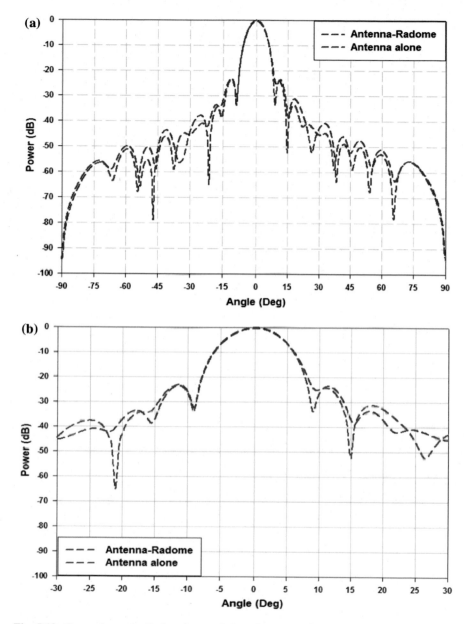

Fig. 5.19 Comparison of radiation characteristics of *antenna alone* and *antenna-radome system* with the seeker antenna orientation at Az = 0°, El = 0° (in the critical nose-cone sector) **a** H-plane radiation patterns, **b** Zoomed plot indicates minimal degradation of the pattern due to radome

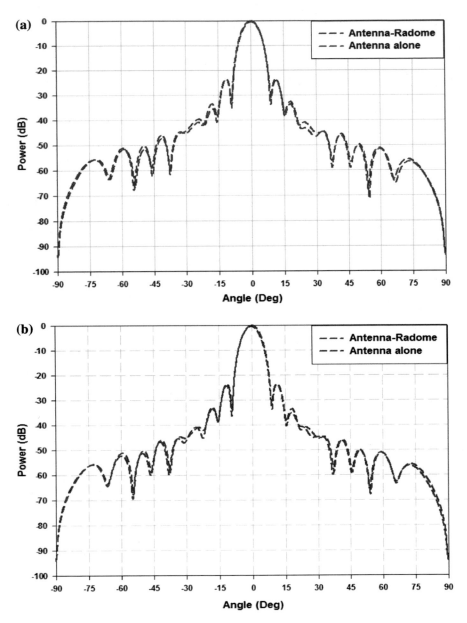

Fig. 5.20 Comparison of radiation characteristics of *antenna alone* and *antenna-radome system* with the seeker antenna orientation at **a** Az = +10°, El = +10°, **b** Az = +10°, El = −10°

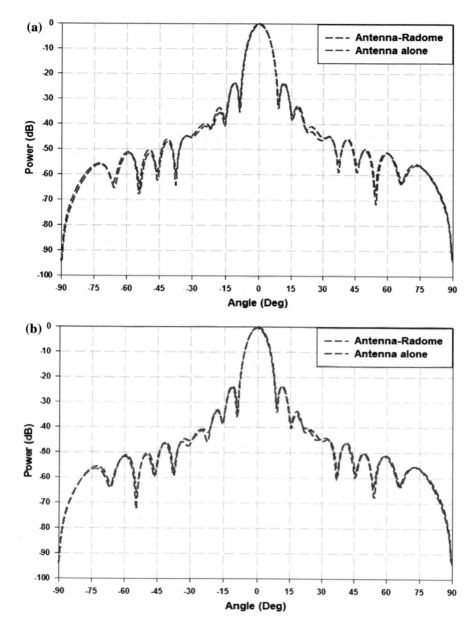

Fig. 5.21 Comparison of radiation characteristics of *antenna alone* and *antenna-radome system* with the seeker antenna orientation at **a** Az = −10°, El = +10°, **b** Az = −10°, El = −10°

angles corresponding to the four quadrants are shown. The computed co-pol shows a gain loss of −0.41 dB at gimbal orientation Az = ±10°, El = +10° and −0.42 dB at Az = ±10°, El = −10° with antenna-radome system as compared to antenna alone system.

Chapter 6
Summary

The present study shows that the graded dielectric inhomogeneous radome exhibits better EM performance characteristics than monolithic half-wave and sandwich radome wall configurations (with either CTR or VTR designs) and provides a bandwidth of 500 MHz from 9.75 to 10.25 GHz with center frequency at 10 GHz. The proposed design gives superior performance for the entire forward hemisphere in terms of co-pol (>-0.5 dB), cross-pol (<-30 dB), boresight error (<3 mrad) and also shows minimal pattern degradation. The excellent EM performance characteristics together with fabrication easiness make graded dielectric radome an exceptional design for both airborne and ground-based radome applications.

© The Author(s) 2018
P. Mahima et al., *Electromagnetic Performance Analysis of Graded Dielectric Inhomogeneous Radomes*, SpringerBriefs in Applied Sciences and Technology, https://doi.org/10.1007/978-981-10-7832-3_6

References

Balanis, C.A. 2005. *Antenna theory, analysis and design*. Hoboken, New Jersey: Wiley. ISBN 0-471-66782-X, 1117 p.

Burks, D.G. 2007. Radomes. In *Antenna engineering handbook*, ed. J.L. Volakis. New York, USA: McGraw-Hill. ISBN 13:978-0-07-147574-7.

Burks, D.G., E.R. Graf, and M.D. Fahey. 1982. A high-frequency analysis of radome-induced radar pointing error. *IEEE Transactions on Antennas and Propagation* 30 (5): 947–955.

Cary, R.H.J. 1983. *The handbook of antenna design*. London: Peter Peregrinus.

Chen, F., Q. Shen, and L. Zhang. 2010. Electromagnetic optimal design and preparation of broadband ceramic radome material with graded porous structure. *Progress in Electromagnetics Research* 105: 445–461.

Huddleston, G.K., H.L. Bassett, and J.M. Newton. 1978. Parametric investigation of radome analysis methods. In *Proceedings of 14th Symposium on Electromagnetic Windows*, Georgia Institute of Technology, Atlanta GA, June 1978.

Kozakoff, D.J. 2010. *Analysis of radome enclosed antennas*. Norwood, USA: Artech House. ISBN 13:978-1-59693-441-2.

Nair, R.U., and R.M. Jha. 2007. Novel A-sandwich radome design for airborne applications. *Electronics Lett.* 43 (15): 787–788.

Nair, R.U., and R.M. Jha. 2009. Electromagnetic performance analysis of a novel monolithic radome for airborne applications. *IEEE Transactions on Antennas and Propagation* 57 (11): 3664–3668.

Nair, R.U., and R.M. Jha. 2014. Electromagnetic design and performance analysis of airborne radomes: Trends and perspectives. *IEEE Antennas and Propagation Magazine* 56 (4): 276–298.

Nair, R.U., S. Sandhya, and R.M. Jha. 2012. Novel inhomogeneous planar layer radome design for airborne applications. *IEEE Antennas and Wireless Propagation Letters* 11: 854–856.

Nair, R.U., D.S. Preethi, and R.M. Jha. 2014. Novel inhomogeneous planar layer radome design for airborne applications. *CMC: Computers Materials & Continua* 40 (2): 131–143.

Nair, R.U., M. Suprava, and R.M. Jha. 2015. Graded dielectric inhomogeneous streamlined radome for airborne applications. *Electronics Letters* 51 (11): 862–863.

© The Author(s) 2018
P. Mahima et al., *Electromagnetic Performance Analysis of Graded Dielectric Inhomogeneous Radomes*, SpringerBriefs in Applied Sciences and Technology,
https://doi.org/10.1007/978-981-10-7832-3

Pei, Y., A. Zeng, L. Zhou, R. Zhang, and K. Xu. 2012. Electromagnetic optimal design for dual-band radome wall with alternating layers of staggered composite and Kagome lattice structure. *Progress in Electromagnetics Research* 122: 437–452.

Siwiak, K., A. Hessel, and L.R. Lewis. 1979. Boresight errors induced by missile radomes. *IEEE Transactions on Antennas and Propagation* 27 (6): 832–841.

Author Index

© The Author(s) 2018
P. Mahima et al., *Electromagnetic Performance Analysis of Graded Dielectric
Inhomogeneous Radomes*, SpringerBriefs in Applied Sciences and Technology,
https://doi.org/10.1007/978-981-10-7832-3

Subject Index

© The Author(s) 2018   49
P. Mahima et al., *Electromagnetic Performance Analysis of Graded Dielectric
Inhomogeneous Radomes*, SpringerBriefs in Applied Sciences and Technology,
https://doi.org/10.1007/978-981-10-7832-3

Monopulse planar array antenna, 15, 17
Monopulse sum channel voltage, 20
Multilayered, 5, 37

N
Nosecone sector, 10

P
Parallel polarization, 9
Permeability, 13
Perpendicular polarization, 8, 9, 22
Planar slab model, 7, 10–12, 23
Power reflection coefficient, 9, 10
Power transmission coefficient, 9, 10

R
Receiving mode approach, 19
Receiving port voltage, 19
Reflection coefficient, 10

S
Seeker antenna, 39, 40
Skin, 3–5, 15, 17
Slotted waveguide, 1, 15, 17

Strength-to-weight ratio, 3, 5

T
Tangent ogive, 10, 15
Taylor, 35–37
TE-pol, 25
Theory of small reflections, 10
Transmission coefficient, 10, 19
Transmission efficiency, 5, 15, 25, 26, 30, 31, 38
Transmitting mode approach, 19
Triangular distribution, 36, 37
Triangular lattice, 15

U
Uniform distribution, 36, 37

V
Variable Thickness Radome (VTR), 1, 15, 21, 29

X
X-band, 10

Printed in the United States
By Bookmasters